STARTING OUT

baby LEOPARDS

I0895304

MEG GREVE

CREATIVE EDUCATION • CREATIVE PAPERBACKS

CONT

ENTS

I AM A CUB.

I am a baby leopard.

Look at my spots. They help me hide in rocks and grass.

I was born in a den with my brother and sister.

Our mom hides us in the tall grass. She moves us around. She keeps us safe.

I get bigger. I get sharp teeth.

My spots turn into rosettes.

I like to play and jump.

Learning skills helps me survive.

I can eat meat when I am two months old. My mom teaches me how to hunt.

I learn to climb and swim too!

SPEAK AND LISTEN

EE-EE!

Can you speak like a cub?

Baby leopards squeak and squeal.

Listen to these sounds:

https://www.pbs.org/wnet/nature/files/2021/04/SoothingBabyLeopardsSocial.mp4?_=1

CUB WORDS

den: a home for cubs in a safe place such as in a cave or under tree roots

hunt: to find and kill animals for food

rosettes: flower-shaped markings on the fur of a leopard

spots: marks or patterns on an animal's fur

READING CORNER

Grady, Tyler. *Leopard: Fascinating Animal Facts for Kids.* Osprey, Fl.: Dylanna Publishing, 2022.

Grodzicki, Jenna. *Baby Leopards.* Minneapolis: Bearcub Books, 2022.

Kellett, Jenny. *The Ultimate Leopard Book for Kids*. Melbourne, Australia: Bellanova Books, 2023.

INDEX

PUBLISHED BY CREATIVE EDUCATION AND CREATIVE PAPERBACKS
P.O. Box 227, Mankato, Minnesota 56002
Creative Education and Creative Paperbacks are imprints of The Creative Company
www.thecreativecompany.us

LIBRARY OF CONGRESS CATALOGING-IN-PUBLICATION DATA
Names: Greve, Meg, author.
Title: Baby leopards / Meg Greve.
Description: Mankato, Minnesota : Creative Education and Creative Paperbacks, [2025] | Series: Starting out | Includes bibliographical references and index. | Audience: Ages 4-7 | Audience: Grades K-1 | Summary: "Introduce beginning readers to the stealthy world of baby leopards with this life science starter. Includes photos, a labeled animal diagram, "Make a Noise" section, glossary, and further resources"-- Provided by publisher.
Identifiers: LCCN 2024014657 (print) | LCCN 2024014658 (ebook) | ISBN 9781640264205 (library binding) | ISBN 9781628329537 (paperback) | ISBN 9781640005846 (ebook)
Subjects: LCSH: Leopard--Juvenile literature. | Leopard--Infancy--Juvenile literature.
Classification: LCC QL737.C23 G748 2025 (print) | LCC QL737.C23 (ebook) | DDC 599.75/541392--dc23/eng/20240419
LC record available at https://lccn.loc.gov/2024014657
LC ebook record available at https://lccn.loc.gov/2024014658

DESIGN AND PRODUCTION
Design by Rhea Magaro
Production by Beeline Media and Design, Inc.
Art direction by Tom Morgan

PHOTOGRAPHS by Alamy Stock Photo/MIKE WALKER, 6, imageBROKER/Paul Sawer, 10; Shutterstock/ Cal Butler, 8, Eric Isselee, cover, 2, 4, 5, 11, 12, 14, Maggy Meyer, 9, Ondrej Prosicky, 7

Printed in China